THE CAMBRIDGE MISCELLANY

XV

THE PROBLEM OF NOISE

THE
PROBLEM OF NOISE

by

F. C. BARTLETT, M.A., F.R.S.

*Professor of Experimental Psychology in
the University of Cambridge*

with a Preface

by

C. S. MYERS, C.B.E., F.R.S.

*Principal of The National Institute of
Industrial Psychology*

CAMBRIDGE

AT THE UNIVERSITY PRESS

1934

CAMBRIDGE UNIVERSITY PRESS
Cambridge, New York, Melbourne, Madrid, Cape Town,
Singapore, São Paulo, Delhi, Tokyo, Mexico City

Cambridge University Press
The Edinburgh Building, Cambridge CB2 8RU, UK

Published in the United States of America by Cambridge University Press, New York

www.cambridge.org
Information on this title: www.cambridge.org/9781107690189

© Cambridge University Press 1934

First published 1934
First paperback edition 2011

A catalogue record for this publication is available from the British Library

ISBN 978-1-107-69018-9 Paperback

CONTENTS

PREFACE

THIS little volume contains, in a somewhat expanded form, the two Heath Clark Lectures which Professor Bartlett delivered recently before a London audience at the invitation of the National Institute of Industrial Psychology. The subject which he chose is one of extreme interest at the moment: newspaper campaigns have been recently conducted and societies have been lately established in order that active measures may be speedily taken to deal with the problem of noise. Not only for reasons of popular interest, but also because of Professor Bartlett's clear, dispassionate and scientific treatment of the matter, his lectures are unquestionably deserving of publication.

As a rule I look with disfavour on a preface written by any one other than the author. But in the present instance it is perhaps only natural that, as Principal of the Institute which arranged these lectures, I should consent to contribute a few introductory words to this helpful and attractive exposition of an important social problem by my former pupil and my successor at Cambridge.

"Opinions" about unwanted sounds, as Professor Bartlett points out, "are rife and sweeping. Facts...are, alas, far more difficult to find." It is facts, not opinions, that the more intelligent section of the community demands about such a debated and debatable subject as the distracting, irritating and harmful effects of noise. In this book the reader will find a concise account of the results of recent reliable investigations of the problem.

He who desires facts will first ask himself—what are we to understand by noise? Professor Bartlett replies, comprehensively enough, that noise may here be regarded as any sound which becomes a nuisance to us.

Next, he will inquire—why and when do sounds become nuisances, and under what conditions and by what mental processes do we become "adapted to" (i.e. are we able to neglect or even to banish from our consciousness) such nuisances?

Finally, he will want to know whether noises have any harmful effect on hearing, under what conditions they produce the greatest distraction and irritation, and whether such distraction, irritation or ultimate "adaptation" affects mental health and mental efficiency, both in normal persons and in the mentally maladjusted or overstrained.

To all these questions Professor Bartlett attempts to reply, so far as our scientific knowledge at the present time permits him to do so. His answers and his attitude will probably satisfy the healthy mind, however unwelcome they may be to the less perfectly balanced.

But even if the ascertained effects of certain noises on the efficiency of every-day work are small, Professor Bartlett wisely recognizes their social importance when they are multiplied by the millions of people who are affected by them. And even if "adaptation" to noise produces apparently no immediate adverse effects, he is careful not to deny the possibility that it may have a remoter influence in subsequent mentality and conduct.

Because the noises to which the average citizen is subjected are hardly ever loud enough even temporarily to impair his hearing; because the effects of an irritating noise on working efficiency are generally small; because we have no evidence that the healthy person suffers from overstrain directly attributable to noise— it does not follow that nothing should be done to reduce noise. A few industrial noises undoubtedly cause deterioration in the operatives' hearing: the ill-effects of unexpected or irregularly recurring noises on highly complex mental work are, admittedly, by no means negligible: not many of us are absolutely free from mental disturbances which in their more pronounced form are termed "psycho-neurotic", and, even when slight, may, sometimes at least, make certain noises distressing.

Professor Bartlett describes in his concluding chapter what can be done to reduce noises and their consequences. As he points out, some noises, e.g. warning noises, are inevitable; but that even warning noises may be needlessly sounded is shown by the striking reduction in the frequency of the hoot of the motor-car horn during the day, which has spontaneously followed its recently enforced abolition at night.

CHARLES S. MYERS

The National Institute of Industrial Psychology
Aldwych House, London, W.C. 2
November 1934

THE PROBLEM OF NOISE

Chapter I

1. Opinions and Facts

Everybody agrees that a lot of noise is a bad thing. Yet each year the world in which we live becomes more and more full of noise. In almost every civilised country expert committees have been appointed to investigate what we all call the "menace of noise". Sporadic newspaper campaigns are carried out and are always popular. They produce masses of letters from masses of people who object to all sorts of sounds which they do not want. Opinions are rife and sweeping. Facts which can be regarded as scientifically established are, alas, far more difficult to find. But the facts about noise are slowly accumulating. Each of the last five years has seen the publication of the results of an increasing number of investigations more or less directly aimed at the problems of this book.

Human reactions to noise in real life are an extremely complicated affair. Science, ignoring perhaps the immediate popular demand, must attack the simpler problems first. A survey of the distribution of topics in recent controlled investigations on noise is not without interest. I have, I think, collected practically all the well-accredited work of the last few years. Of this 60 per cent. is either entirely or predominantly physical, the bulk of it dealing with most fundamental and necessary attempts to achieve a measurement of noise; 28 per cent. deals with the direct physiological effects of noise stimuli, and of the remaining 12 per cent. only a very few papers discuss those psychological problems which, for the ordinary person, whether he is engaged in heavy industry or not, constitute the most important and interesting of the questions connected with unwanted auditory stimulation. Precisely as should be expected, perhaps, it is just where opinions are the most common and the most dogmatic that the established facts are the most scanty.

2. WHAT IS NOISE?

Noise is any sound which is treated as a nuisance. This neither appears nor is a very exact definition; but, from the point of view of the present discussion, it is probably as good a one as can be obtained. Certainly the physical definition of noise as sound resulting from stimuli which cannot be resolved into periodic vibrations is, apart from other difficulties[1], a hopeless one. At the basement of the laboratory in which I work, an electrical generating plant is in constant use. The dynamos produce a musical tone which can be heard all over the building. I have yet to meet any research worker in the laboratory who fails to treat this musical tone as a noise. Even a tuning-fork, yielding only the purest tone, can be a horrid nuisance on occasion. Let anyone sit in a sound-proof room—as I have repeatedly done—for two hours at a stretch and listen to the sound of an electrically maintained tuning-fork coming intermittently throughout the whole of that period. At the end of the experiment he may feel inclined to use much the same language of the tuning-fork as that which the less restrained members of the community employ when they write to newspapers about clattering milk pans, clanging bells, or shattering motorhorns.

If we turn to physiology for a definition we get at the moment less help still. The physiologist must attempt to define noise in terms of the reacting mechanisms in the auditory apparatus of the body. Broadly speaking, as most people know, the human ear contains three working sections. The outer ear collects and directs the sound waves which come to us from their external source; the middle ear transmits these waves to the inner ear, and it is from the delicate and intricate structure of the inner ear that the impulses are initiated which pass along the auditory nerves to the brain, and so make it possible for us to identify and interpret the sounds which have reached us. Nobody knows precisely how these physiological mechanisms work in the case of noise. Their most important secrets lie undoubtedly in the activities of the inner ear, of the auditory nerve tracts, and of the end stations in the brain.

Recent research seems to point unmistakably to the cochlea—the inner ear—as acting in some respects as a resonator and in some respects as a telephone(2). But how, in either respect, it distinguishes tone from noise—if indeed it does this—is wholly unknown. Moreover, that aspect of noise which is the most important for us, namely the annoyance which it produces, certainly cannot have a location in the ear, and as certainly nobody is within measurable distance of knowing whether it has, or where it has, a place in the brain.

We must therefore fall back on the rough working definition that any sound is a noise when it is treated as a nuisance. Why and when are sounds treated as a nuisance? It is useful, though not, of course, decisive, to attempt some preliminary classification of reasons. One guiding fact must be held in mind all through. The qualities of any particular sound depend largely upon the background against which the sound is experienced. This background may be auditory or not. If a man hears an unusual sound coming apparently from the inside of his car when he is driving, he is likely to be disturbed and irritated until he can identify and localise it, although a very similar sound with a different background might pass unnoticed or produce no annoyance. In a large number of cases, among them many which may be of great importance in industry, the background is not auditory at all. When we get rid of an unwanted bore, the slam of the door as he disappears may be grateful to the ears, though door slamming may take a high place in our catalogue of nuisances under other conditions. To this important question of noise relation to its background I shall return more seriously later.

There are certain characteristics of sounds which make them stand out, or "attract attention", on almost any background. The most important of these are loudness, ambiguity of direction and unfamiliarity. Of these the only one with an obvious physical correlate which seems capable at present of being dealt with in a reasonably accurate manner is loudness; and so it is not surprising that loudness and noise are closely connected in the popular mind, or that a large part of recent research should have been concerned almost wholly with this quality.

Now it is obvious that to be able to state the relative position of a sound in a scale of intensities itself throws no light whatever upon the effects of noise on work. Further, in treating loudness as a correlate of intensity, we are clearly isolating one characteristic only of a number which noisy sounds possess. The characteristic is doubtless important, but we must not conclude at once that because it has received more detailed treatment than any other it is of necessity the most important. Still, if we are to arrive at any safe generalisations about the effects of noise on work, it looks as if we must be able to compare noises in some way, and even a partial and inaccurate way is better than none at all.

3. THE MEASUREMENT OF NOISE

We must therefore turn first to recent researches into what is called the "measurement of noise". By measurement of noise is generally meant the equation of a sound to a standard sound in respect to one or more of its characteristics— loudness is almost the only characteristic that is actually used—when the position of the standard sound in a scale of intensities is known.

Broadly speaking, two methods are possible. The first is a purely physical method, in which an attempt is made to measure the actual energy output of a given source of noise and to equate it with a similarly measured output of a standard sound source. This, from our present point of view, is of very little use, not only because the apparatus required is delicate, not easily portable and requires expert management, but because, even when the energy output of every component of a complex noise is known, we are still very far from being able to conclude anything about its total effect when all the constituents are simultaneously present. (But see also § 7 of the present chapter.) The masking and interfering effects of constituents of different frequency might even mean that a reduction of the energy output in the case of some of the constituents would leave the total effect more thrusting and annoying than it was before.

Consequently, both for theoretical and practical reasons, the method almost always adopted is a mixture of a physical and a psychological procedure: the measurement is an aural measurement. All the devices in use derive from a scheme originally proposed by Barkhausen. A vibrating reed or tuning-fork is used as a standard source of sound. The stimuli are conveyed to a telephone which is held near or against the observer's ear. Then the standard sound is adjusted in intensity, by the variation of a calibrated potentiometer, until it appears to be just as loud as a noise heard by the other ear, and the noise is expressed in whatever units of intensity are being used to indicate the position of the standard sound on a scale of intensities[3].

At first sight this looks as if it must be a very inaccurate method, admitting wide divergencies of judgment. Yet a considerable number of independent investigators, in this and other countries, have demonstrated that with trained observers the method yields results which are unexpectedly consistent and of practical value[4].

The Bell Telephone Company of America have developed a slightly different technique, a little less easy to use, perhaps, but particularly adapted to certain situations[5]. With this the telephone which conveys the standard sound is so held that the noise and the comparison tone both enter the same ear. The normal threshold for the standard sound—that is to say the minimum audible note—is known. Now the standard tone has to be adjusted until it is just drowned, or masked, by the noise; and the noise

is expressed in terms of the rise in threshold*
which its presence produces. This method is obviously the better one whenever a task is set
which demands the identification of a particular
sound upon a background of noise.

The simple method suggested and used with
great success by A. H. Davis in this country
combines both of those already mentioned(6).
With this a standard tuning-fork is used. The
fork is struck and held in a constant position
near the ear of the observer. The rate at which its
sound decays is known, and the tone can be either
equated with the apparent loudness of a given
noise, or the moment at which the fork ceases to
be heard when the noise is present can be recorded.

These are the methods. In what sense do they
measure noise? Clearly noise is not being measured directly. It is being equated, for one only

* "Threshold" is defined as the minimum
value of a stimulus which is capable of producing
a sensory response. Sometimes the value is as-
signed which produces 50 per cent. correct reactions in a series of experimentally controlled
tests; more often, and probably more satis-
factorily, 80 per cent. correct reactions are
required.

of its many characteristics, with a standard sound of a particular frequency. The story of the introduction and rapid spread of the decibel as a unit to express the intensity level of a sound stimulus has been told many times and need not be repeated here(7). Speaking very approximately indeed, when we say that the noise of a pneumatic drill at 20 ft. distance is round about 90 db. above threshold, *all* that we mean is that it appears equal in loudness to the sound of a tuning-fork of 640 cycles a second*, when that sound has been increased by about ninety steps of perceptible difference above its minimum audible value. We have said nothing directly about the loudness of the pneumatic drill. The sound which it produces is a combination of many different frequencies, and in strictness

* An observer may, of course, use as his standard sound a tuning-fork of any desired pitch. Davis chose the frequency of 640 because this lies well within the region of high sensitivity for the normal human ear, and produces a sound variations in the loudness of which are easily and consistently appreciated. But some observers have preferred to work with standard sounds of higher pitch.

each component has its slightly different decibel scale. Similarly we can say that the sound of a powerful motor-horn at about 60 ft. distance is some 90 db. above threshold. It does not follow that the noise of the horn would necessarily sound as loud as, and not louder than, that of the drill if they were being directly compared. In precisely the same way we could make a standard salt solution and equate the saltiness of bloaters and that of olives to this. But we still could not at once express the saltiness of the bloaters in terms of the saltiness of the olives. What we could say, however, and this would be a very valuable thing to know, is that when the saltiness of any compound occupies a high position in relation to a standard scale it tends to become disagreeable. And we can say of noises that when, compared with a standard sound of known frequency, they are equated with a high value of intensity level, they tend to become disagreeable, or harmful, or both. More important still for many practical purposes, we can begin to say, if we adopt the masking method of measurement, that when certain noises are present, and it is at the same time necessary to react meaningfully to other sounds, this will present irritating difficulties when the noises, as compared with the standard sound, are at a high level of relative intensity.

In the light of these methods let us turn to recent work on the direct effects of the loudness of sounds upon human auditory responses.

4. THE EFFECTS OF NOISE ON HEARING

First we must consider the effect of the loudness of noise on hearing itself. Recent work, especially by Rawdon Smith at Cambridge[8], has demonstrated that sounds of about 900–1100 cycles a second must be increased to the region of 100 db. above the threshold of hearing if they are to leave any marked and measurable after-effects on auditory acuity. Even then the effects are generally swiftly over, though for a few seconds the threshold may be raised by an amount equivalent to some 20 db. on the intensity scale for this frequency. Lesser amplification will produce similar results in the case of sounds of higher frequency and a considerably greater amplification is required to produce equivalent results with sounds of lower frequency.

13

The transition from tones to the irritating sounds of everyday life must be made with great caution. The latter are immensely more complex than the former. Yet repeated observations fit well with the experimental work. It is only in highly selected and special occupations that there is any evidence of serious damage to hearing done by noise. Airmans' deafness is reported frequently, and may apparently last for some considerable time after a flight (9). Unfortunately either it has never been systematically studied, or the results of such study have not been made public. The airman, while he is flying, lives in a noisy environment which, measured by the standard methods, often reaches and surpasses the intensity level represented by 110–120 db.

Tube-train drivers, and even occasionally passengers, report temporary deafness sometimes, and here, again, the normal intensity level of noise may reach 90–100 db.

Boiler-makers' deafness is well known and has been extensively studied by clinical and experimental methods (10). The boiler-maker at work, especially when he is inside the boiler, may have to endure a loudness considerably greater than

14

anything used in experiments upon human beings. Not only so, but many of the constituents of the noise are of a high frequency, so that their loudness effect is greatly amplified. That continued exposure to such conditions may produce inner ear inflammation and lesions and lead to permanent deafness seems certain. Wittmaack and Yoshii, working independently, have conducted controlled experiments with guinea-pigs under conditions parallel to those of boiler-making and have found inner ear inflammation —not, apparently, very well localised—in a number of cases[11]. Some other industries may produce more or less similar effects. N. Losanoff, working in a nail factory at Saratov, and using rabbits and dogs as experimental controls, found inner ear lesions set up by the noise, the relative intensity level of the sounds not being reported in this case. He also found, as might be expected, that any conditions promoting mouth-breathing tended to intensify the effects[12].

All these are special cases, however, and can be dealt with by the use of particular defence devices which are outside our purview at the moment. The loudness of noise which the average citizen has to bear hardly ever approaches anything that has been shown to leave harmful effects upon acuity of hearing. And when it does, the noise is practically always transient and intermittent. It can be concluded with certainty that so far as the normal person is concerned, unless he is occupied in certain highly specific tasks, no convincing case can be made out against noise on the ground that it is likely to produce either permanent or transitory loss of efficiency in his hearing.

5. The Effects of Noise on other Physiological Reactions

What about other possible direct physiological effects of the loudness of sounds? R. Nyssen and J. Helmsmoortel, Jr., report that intense auditory excitation directly increases arterial pressure, especially in the systolic phase[13]. E. L. Smith and D. A. Laird found that in four healthy observers a noise equivalent to between 80 and 90 db. intensity level decreased normal stomach contractions by 37 per cent., and thought this evidence that the sounds were setting up something like a fear response[14].

16

F. L. Harmon has recently reported that unexpected and unfamiliar noise, introduced during mental work, may increase metabolic rate, heart rate, breathing rate and body volume by as much as 67 per cent.(15) But all these results are transient. They are incident to a phase of maladaptation to the noise, and in a normal subject they quickly pass away and do not return.

We can sum up: as regards loudness—

(*a*) Noises which, measured as already indicated, are equivalent to an intensity level of anything above 50–60 db. are likely to attract and control attention.

(*b*) For frequencies of 900–1100 cycles a second, intensities of about, or above, 100 db. produce marked after-effects and temporarily or permanently impair hearing. For frequencies above this range the requisite intensity level is less; for frequencies below this range the requisite intensity level is greater.

(*c*) Except in selected occupations the normal person is very rarely indeed submitted to sufficiently persistent intense noise to produce any permanent effect whatever upon auditory efficiency.

(*d*) The other physiological effects of noise which have been demonstrated are incident to a phase of maladaptation and normally pass away rapidly and completely.

At the same time there may be supersensitive individuals for whom these generalisations do not hold. But if so they are certainly much more rare than is often supposed. There are also individuals who have a history of ear trouble originally set up by causes that had nothing to do with noise. If these are submitted to heavy noise for long periods, even although the noise level does not reach that which would seriously affect a normal person, the old troubles may return in an aggravated form. But so far as I know there is no exact knowledge concerning this,* and no satisfactory statistical survey. F. H. Frankel, working in nine heavy industrial plants in New York, found very little evidence of impairment of hearing without earlier history of bad ear

* A. W. G. Ewing, in an important paper on "Auditory Fatigue" (*Brit. Journ. Psychol.* xxv, 3), has made some study of the effects of persistent sound stimulation upon two observers suffering from middle ear disease.

conditions(16). At the same time the greatest percentage of hearing deterioration was found in the groups subjected to the greatest noise. Here is a topic not at all difficult to investigate which certainly would repay further research.

6. The Effects of Ambiguity of Direction and Unfamiliarity

We saw that there are broadly three characteristics of sound which are thrusting or dominant in almost any background. Loudness has now been considered; the other two were ambiguity of direction and unfamiliarity. When I was preparing this book, my boy of eleven, who is extremely interested in any sound emitted by machines, began to talk about the matter. I said:

"Do you find noises very disturbing; when you are in school, for instance?"

"Some are, and some aren't", he answered, and then added decisively, "it depends on whether you know where they're coming from."

The observation is accurate. Special difficulty in the location of the source of a sound is sure to arise with noises containing predominant com-

19

ponents of very high or of very low pitch, and this undoubtedly accounts in part for the common observation that such noises are apt to attract more than their due share of attention. Further, sounds coming from above the observer are, on the whole, particularly difficult to locate accurately without the help of special devices, and this provides a part of the explanation of the fact that people in general are very ready to stop whatever they may be doing and gaze upwards whenever a noise comes from above them, until they are satisfied that they have identified its source. A great deal is known about the auditory basis of the location of sound(17), but very little about the relation of the directional quality of noise to its capacity for producing distraction and disturbance. A much more thorough investigation, for instance, than has ever yet been made upon the effects on animal orientation and general behaviour of noises coming from different directions, ought to yield results not only of theoretical interest, but also of practical significance*.

* Mr A. S. E. Ackerman, on the basis of a great amount of practical experience, points out

Although lack of directional quality and unfamiliarity—characteristics which obviously overlap, but as clearly are not identical—might readily be made the subject of controlled experiment in relation to their effects upon work, I do not think they have been. And in any case the effects, whatever they may be, are most emphatically not to be discussed in terms of peripheral mechanisms. They involve processes of a

to me that he has found that "people are more often annoyed by noises which they know are produced near to them, as just the other side of a wall, rather than by the same quantity of noise arriving from a distant and greater source". There can be no doubt that this is very frequently the case, especially when the source of the sound, though near, cannot actually be seen. The comment, in fact, raises the extremely interesting point of the interplay between ears and eyes in regard to the distracting qualities of sound. In general when a person can *see* what is producing a noise, provided the sounds in themselves are not especially interesting, or difficult to interpret, adaptation is quicker and more complete. Some important theoretical points are involved here which would well repay more exact investigation.

21

definitely psychological order, and to these we shall turn in the second chapter.

7. Recording the Characters of Complex Sounds

Many devices have been proposed for recording, photographically or otherwise, the various component characteristics of complex sounds and their relative loudness. A graphical record has certain obvious advantages. Once obtained it remains and can be analysed at leisure, whereas so far as the actual auditory perception of the sound goes, when the stimulus ceases the sound can no longer be accurately dealt with. Further than that, most people who are not specially trained find the analysis and interpretation of a visual record much easier than direct auditory analysis. But one of the difficulties in the way of the production of practically useful graphical records of complex sound stimuli is that the human ear, as a receptor organ, differs in important respects from most of the physical receptor devices that have been used.

The sensitivity of the human ear varies very considerably with the pitch of the sound which is being received. It is low for relatively low and also for very high tones, and hence loudness as objectively measured or recorded may fail to give a true picture of the actual perception of noises which have many components. Further, the ear mechanism itself produces some distortion of the sound waves as they reach it, and this may give rise to certain auditory phenomena which find no representation in a purely objective record. And again, when a complex noise is received by the ear, certain of its components, owing to the selective sensitivity of the hearing mechanism, may be masked or drowned by others, with the result that some tones which are objectively recorded fail to be heard altogether.

However, with the brilliant development, during the last few years, of electrical methods for the transmission, amplification and recording of sound—a development consequent upon the enormous spread of wireless transmission and broadcasting—it is now possible to overcome most of these difficulties. By means of a modern transmitter, a system of amplifiers and an oscillograph, it is possible to record for permanent use very complex noise forms approximately as they actually make their impression on the hearing mechanisms.

The uses of a recording apparatus possessing an ear-like frequency-response are obvious. It is, for example, possible now to record graphically the explosive noise made by an internal combustion engine both without and with a silencer, not merely depicting faithfully the actual objective pressure of each component of the complex sound, but showing them in such a way that the relative audibilities of the different constituents take their place in the whole as they are actually perceived. With such apparatus the efficiency of various devices proposed for the reduction of noise can be accurately studied, and problems of sound absorption and transmission stand a far better chance than ever before of being solved in a practical manner. Further, the very important group of practical questions involved in the phenomena known as the "masking" of sounds can be understood in a way that would, only a year or two since, have been totally impossible(18).

8. THE MASKING OF SOUND

Closely related to studies of the loudness of sounds and to our central problems is another matter at which we must glance. The pneumatic drill, the hoot of the motor-horn, the rattle of a busy city street may actually deafen nobody. But they may cause us to miss what our friend is saying, or to overlook the chiming of the clock that tells us that lunch time is drawing near, or that a lecture is drawing to its end. They continually mask sounds more important than themselves, and so, rightly, we feel irritated and annoyed. Into the details of the study of the masking effects of sounds, a study which, largely under the stimulating influence of Harvey Fletcher[19], has made striking progress in recent times, it is not possible to enter. But the guiding facts can be expressed briefly. In general, masking to any marked degree occurs only when the frequencies concerned are neither very widely separated, nor are so related as to produce beats. Thus a very high tone will ring through much more intense low tones—as we always hear the piccolo in an orchestra, even when the big brass instruments are sending out a great rush of sound—or, again, a low tone may be masked by a higher tone which is sufficiently intense and is not beating with it.

Most of the sounds that are treated as noise in everyday life are complex. They contain components of different frequency, and the loudness levels of all of these may vary among themselves. Thus some of the components may mask, or drown others, and the noise as a whole is heard as possessing a certain auditory quality. But if the sound is now amplified, so that each component frequency undergoes magnification to about the same absolute degree, the masking effects of the different components may be very greatly altered. Some of the frequencies which perhaps were not heard at all before become easily perceptible, while some others, which before were prominent, may cease to produce any, or any noticeable, effects. In the result the noise may appear to be very different. Amplification can, in fact, be so arranged as to make a pleasing sound more displeasing, or a displeasing sound more pleasing.

Very nearly everybody has now had plenty of opportunity to observe these phenomena, sometimes to his pleasure, more often, perhaps, to his annoyance, in loud-speakers, where amplification has altered greatly the apparent quality of the sounds reproduced. It follows that if a task has to be carried out which requires an observer to detect and interpret a given sound against a background of noise, and if either the special sound or its background is controllable, much worry and bad work may be avoided when either the one or the other can be so adjusted that masking effects are at a minimum; but naturally every case has to be considered in relation to its own immediate conditions.

I have already remarked that a low tone has in general very little masking effect on a tone of considerably higher frequency. It has, for instance, often been noticed that women operatives, with their higher-pitched voices, can usually converse much more easily in a noisy workroom than can men. And of course this means that wherever certain sounds are employed as warnings, and have frequently to be brought into use in a noisy environment, these sounds will almost always be found to contain strong components of high pitch. Now it so happens—very likely largely because they are particularly difficult to mask—that high tones are very productive of annoyance. From this, practical problems are obviously bound to arise. The warning sounds cannot properly perform their desirable functions unless they are high in pitch. If they are high in pitch, they will also acquire other undesirable functions. There is no easy way out of the difficulty. But such warning sounds do not, as a rule, need to last very long. When, by the thoughtless or inconsiderate, they are used more frequently than is necessary, they may however become a most pressing nuisance.

In this whole realm, once again, we approach the more complex psychological problems. Masking is a physical and physiological effect, but its reaction upon work is by way of the psychological mechanism. It operates by slowly piling up annoyance, irritation and feeling of defeat. We can bear with equanimity the masking of many sounds, but when the sounds that are distorted, or prevented from coming through, are those upon which our most important interests turn, the masking noise becomes a real enemy. What happens in such situations is our next subject for study.

Chapter II

1. PSYCHOLOGICAL PROBLEMS

The first of these chapters may perhaps be re-garded as a clearing of the stage for action. I de-scribed briefly certain fundamental researches, largely physical in nature, no doubt necessary if noise is to be treated in a comparative manner and conclusions reached which are not to be considered as merely relative to particular and immediate circumstances; but in themselves throwing no light upon the effects of noise other than those upon auditory acuity itself. We then found that the intensity level necessary to pro-duce disturbances of hearing is something much higher than the average citizen ever has to meet, except, perhaps, for a moment or two at a time, in his everyday life. This, however, need worry nobody. We are all convinced that noise is an enemy, and that its main attack is not directly physiological, but by weapons of an insidious and subtle psychological order. It is upon the psychologist that the task of marshalling the evidence against noise devolves. Now therefore let us prepare for the stories of broken nights and broken nerves, of shattering headaches and irritable tempers. And should some psychologists, emulating, maybe, the methods of more remote sciences, try a few experiments and attempt to tell us that some of these heart-rending stories have less foundation than is imagined, let us not take too much notice of them. For surely we should not enjoy our civilisation half as much as we do unless we could be frequently pointing out how thoroughly bad it is.

There are two groups of psychological problems involved. First there is the question of the effects of noise upon what the worker does, and secondly there is its influence upon the worker himself and his experience. Some people would confine the psychologist's attention to the second of these groups, but from this view I dissent. The psychologist has to study the determination and control of any form of human conduct, and he is interested in experience just in so far as this helps to decide how a person behaves. What then of the work?

2. PSYCHOLOGICAL EXPERIMENTS ON NOISE

In bulk the experimental work of a psychological order dealing with the effects of noise is not at present impressive; in unanimity of results on the whole it is. D. A. Laird shut up four healthy typists in a specially constructed test chamber[20]. He made them work in "noisy" conditions and with reduction of noise. The level of noise is reported as "50 units". It is not clear what the units were, but I suppose the noise was measured in one of the ways indicated in the first chapter. In the reduction periods the level of the noise was cut down to "40 units". We can thus say that, although the sound stimuli were far below an intensity required to affect hearing, they were certainly of the order that normally attracts and may control attention. The quality of work was scarcely affected by noise, the quantity of work decreased a little in the case of the two more expert typists when the higher level of noise was present. Nothing is here for tears.

H. M. Vernon carried out a number of experiments which, I think, have hardly received the recognition they deserve[21]. He, aided by a brave and devoted assistant, carried out a variety of mental tasks in a variety of noisy circumstances. To get some indication of the general lie of the problems, he first used fairly controlled sounds: loud, medium and soft music, a rapidly ticking metronome, and a combination of mixed noises presented irregularly. These are, of course, noise only in the sense in which the term was defined in the first chapter. Some of them were not even that, for on the whole the musical background was welcomed. In no case was there any noticeable effect on the accuracy or on the speed of the work done, whether the work was mental or manual. He therefore transferred the scene of his operations to three large factories. The intensity level of the noise under these conditions is not reported, but seeing that, in every case, conversation had to be shouted this must have been sufficiently high.

Dissatisfied with the meagre effects shown in the objective records, Vernon now hit upon the plan of attempting to register the degree of subjective disturbance produced by the noise. This was not a new method. Attempts had already been made, at which we will glance in a moment or two, to work out annoyance curves for loud sounds. In the more or less parallel case of the disturbing effects produced by glaring light, subjective estimates of the amount of disturbance had been made the basis of conclusions about street lighting, office illumination and head-lamps on motor-cars. But Vernon made a more systematic study than had been attempted before with this method of subjective estimates in relation to problems of noise. Now the justification of the method is exactly the same as that of the methods of measuring noise which we considered in the first chapter. With trained and careful observers, the method does yield unexpectedly consistent results, in so far as any particular person is concerned. When the physicist who is trying to measure noise points to this consistency of judgment as a justification of his method we tend, on the whole, to acquiesce, perhaps even with enthusiasm. When a psychologist uses the same criterion of consistency in the same manner, for some reason or other we still tend to be critical.

At the risk, therefore, of a brief digression, it is worth while considering what the unfortunate psychologist is to do when he is asked, as he very frequently is, to investigate the way in which human behaviour is affected by abnormal conditions of work. At first sight the problem looks a perfectly easy one. The psychologist must make his observer carry out some simple task, the amount or the quality of which can be readily assessed. This must be done under ordinary, controlled conditions. Then the abnormal conditions can be brought in, and any change in the quantity or quality of the performance recorded. On paper the scheme looks admirable; in fact it is beset with difficulties, and here are some of them:

(1) Normally everybody works at any task a good long way below his maximum efficiency. If we did not, when the course of life made an extra demand we should crack and collapse. Even when we honestly think and say that we are putting out our utmost efforts, we very rarely indeed are doing so. Thus when the psychologist brings in his abnormal conditions, all that happens is that the observer makes a little extra effort, and may not even know that he is doing so, and the work proceeds as well as, or

even better than, before. Of course it may be that the small additional effort piles up and eventually produces an intolerable strain; but, as we shall soon see, there are powerful factors which can counteract this effect.

(2) Any simple task, the performance of which is easily measurable, also can be rapidly automatised. Now it is a bit of everyday psychology—often, however, forgotten—that the more mechanical a task becomes, the more it subsides into habit, the more is it dissociated from *any* particular background. All our ordinary daily actions are performed to much the same effect, although the daily external circumstances vary very widely. And so, when the abnormal conditions are imposed, the task goes on as before, not because the abnormal conditions can have no effect, but merely because the work has become so much a matter of routine that it is cut away from any special conditions. Even if the archangel Gabriel should fill the world with sound, there are plenty of people who would brush their hair, put on their clothes, eat their food, raise their hats to a lady, exactly as usual. Psychologists have shown a perfectly

amazing ingenuity and fertility in adapting or inventing simple tasks to aid them in their attempts to study the effects upon behaviour of unusual conditions. The results are, just as Vernon states them to be, almost invariably disappointing.

Very well then, why not run to the other extreme and experiment with a really complex and difficult task? For here we might get something which is on a more delicate balance so to speak, and is thus the more likely easily to be tipped over one way or another by anything unusual in the environment. If we do this we run into other difficulties.

(3) It is true that a complex performance will improve with practice for unexpectedly long stretches of time, and that the positive practice effects may mask any negative effects of unfavourable external conditions; but there are ways of circumventing this. It is true that the more complex the performance the more difficult it is to measure it in any exact manner; but there are ways of circumventing this. The real difficulty lies elsewhere. The more complex and difficult a task, the more absorbing to a normal

subject it is likely to be. The more interesting and absorbing it is, the more the performer is likely automatically to cut out all external circumstances. Take the matter of noise itself. Professor Douglas Fryer, working at Cambridge, began to investigate the effects of different patterns of thrusting sounds upon certain sorts of mental work (22). He used an ingenious arithmetical test, so arranged that all of his observers got extremely interested, mainly because they wanted intensely to beat one another, or to surpass some previous performance of their own. For his noise stimuli he used a set of clanging bells, bells being, as everybody knows, a very common cause of popular complaint. There could be no manner of doubt that the bells, whose pattern of stimulation could be changed at the experimenter's will, were genuinely disturbing. People working in other rooms all over the building were at first moved to make unflattering remarks. But what happened in the experimental room itself? The observers who were working at the test, after a day or so, simply shut the noise out, treated it as not a part of the experiment at all, and carried on, apparently without strain and with no increase of inefficiency. They had to be told again and again that the noise pattern must be treated as essential to the test performances.

There are, then, many difficulties with the complex type of task. But all the same this does offer a more promising approach than does the simple performance, and I think it safe to say that future experiments will be found to yield more convincing results along this line of technique than along any other.

There is one further possibility. If we could arrange work so that it must be carried out at a speed and with an efficiency near the observer's maximum, we might expect the effect of abnormal conditions to be more delicately mirrored in the performance. We have recently adopted this method with considerable success in the parallel case of glare [23]. It is less easy to arrange the conditions in the case of noise so that they would have obvious significance for ordinary conditions of life, but it is not too much to hope that an adequate method will be found.

Vernon was well aware of all these difficulties of measuring the effects of noise in terms of performance. He proposed that he and his assistant should instead try to assess the degree of distraction produced by various sounds. He used five categories of disturbance: none, slight, moderate, considerable and very considerable. The loud and swift ticking metronome produced at first moderate or considerable disturbance, provided a fairly complex mental task was required of the observer. An intense shaking noise, made by machines in the bar-chocolate department of a large chocolate factory, effected considerable disturbance, shading off, after an hour or so, to slight or less than slight distraction. Almost exactly the same course of subjective experience was followed in a large nail factory, and, with somewhat less emphasis, in the clerical department of another large industrial concern.

Two important facts emerge:

(1) The disturbing or distracting effect of sound, when an observer is engaged upon some special task, is not a function of the noise itself, but of the noise in relation to the task. On the whole the more the work puts a demand on the higher mental processes, the more disturbing is the noise likely to be.

(2) But the distracting effects of noise tend to diminish rapidly, provided the noise is tolerably continuous and regular. If it is discontinuous, coming and going at unexpected moments, the disturbance is both greater and more prolonged.

We can also say two other things: the first is that, on the whole, noise, however intense it is, is not very likely to disturb the people who make it or even the people whose exposure to it is regular and long continued; and the second is that probably the disturbing effects of noise are · at their maximum for people who have to do mental work, but are for some reason bored, tired, forced into a job which is a bit too difficult for them or not quite difficult enough, or in which they are only moderately interested. If any community contains a considerable number of people of whom these things are true, complaints of noise will be common.

As I have already mentioned, annoyance curves for loud sounds have been worked out. They must be taken as approximate only, but so far as they go they are definite enough. Laird and Coye show that for notes up to about 500 cycles a second the degree of annoyance follows fairly closely the loudness of the tone (24). Above that frequency the annoyance produced tends to increase in more rapid proportion as loudness increases; it is as if, somehow, in the higher frequencies, pitch itself is inherently distracting.

These results, interesting as they are, leave one question untouched. We assume that anything which noticeably annoys us must have an adverse effect upon our work. This, however, so far as the experiments considered go, is not proved, and it certainly must not be taken for granted.

I now come to some experiments directly concerned with the psychological effects of noise carried out at Cambridge. For these I was myself largely responsible. The work was made possible by the Medical Research Council and the report was given wide publicity when it appeared (25). It was given almost as wide disapprobation. I had better, therefore, describe briefly what we did and what results we regard as established.

We worked first in a very noisy room in a printing press. We did this mainly in order to discover what technique of experiment would be best calculated to bring out the effects of noise on performance. When we had got our methods we transferred to the laboratory, where the noise stimulation could be presented in a relatively controlled manner. We used a variety of noises, some of them, such as loud clicks, electrically transmitted to the ears, or an imitation of a noisy workshop, were "noise" in the common sense of the word; and some of them, such as loud and soft gramophone records, were "noise" only in the sense defined in my first chapter. We did not report the loudness level of the sounds used, but in fact it ranged from some 40 db. above threshold at the lower limit—except in the case of the soft gramophone—to some 80–90 db. at the upper limit. That is to say, at its least it was loud enough to command attention, at its most it was below the range likely to impair hearing. The performances required varied from very simple manual tasks, through a slightly more complex combined manual and mental task, consisting of the assembly of a piece of apparatus made up of 14 units, to definitely complex mental work requiring the solution of problems. There were in all seven different groups of experiments and about 130 observers. All the experiments were, of course, adequately controlled.

In the case of each investigation conducted, the presence of noise was followed by some initial drop, either in the quality, or in the quantity of work done, or in both. This did not happen with every observer in any experimental group, but it held true of the majority of observers in all groups. The deterioration was not great, and in fact, if we take any single group of experiments it was not, save in one or two cases, statistically significant. It was there, however, small in amount, but consistent in direction. Other things being equal, the louder noise produced the greater effect, but discontinuous or intermittent noise affected performance more than relatively constant noise, even when the latter was of greater intensity. Further, quiet noise, the soft gramophone for instance, was definitely more disturbing when it was interesting, but its precise significance was not clearly audible. Cases of intense irritation and annoyance with the noises were common in early stages of exposure. The irritation always passed away, after a few minutes in some cases, an hour or two in others, a day or two in others. Irritation or annoyance was found to be perfectly consistent with a high level of performance. In our cases, as in all the others that have been experimentally investigated, the apparently harmful effects of noise, whether objective or subjective, tended to disappear rapidly as our observers got used to the conditions of their work.

These were the results generally expressed. They were widely criticised, I think mainly on three grounds. First all our observers were normal and healthy young people. Apparently, for some reason or other, we ought to have emulated the man in the parable. We should have gone into the highways and byways and compelled, not, indeed, the halt, the maimed and the blind, but the tired, the bored and the fretful to come in. I still can see no good reason for this. I should be sorry to think that in such fashion we might have got a better sample of the population of a contemporary community than we did.

Secondly, our experiments were not sufficiently prolonged. There are certain obvious difficulties in the undue prolongation of any experiment of this type. But these apart, we are here certainly in the realm of theory. It appears that when the noise stimuli first break in they adversely affect performance. Now we can assume, if we like to do so, that when the obvious bad effects disappear, the noise continues to be as harmful as ever, but the observer, by a little extra effort, is counteracting the tendency to do less, or to do less well. Then there will be minor, but slowly accumulating strain, and what we call getting used to the sound is really a continuous exercise of control which is very likely eventually to break out into some form of irritation, or headache, or bad temper, or something of that sort. That may be the case, but it is wholly wrong to assume that it must be the case. An alternative is possible. The noise may become definitely accepted, assimilated as a part of the working background of the task. Literally the pattern of sensory stimulation then changes. The cues to work have become different. There is no necessary, underlying, sub-conscious conflict, and so there is no necessary slow piling up

45

of harmful effect. We preferred this second alternative in the case of our own subjects, because there was nothing in their behaviour, even after relatively long periods of exposure to noise, to warrant the first. Moreover, it never has been demonstrated that a normally healthy person, submitted to prolonged stimulation by noises of a level of intensity or of disagreeableness commonly met with in everyday life, and doing tasks of the degree of difficulty commonly demanded in everyday life, shows any unusual fatigue or strain that can be traced directly to the noise. Unusual fatigue or strain do appear in plenty of cases, but then there are usually—I should myself say always—other important factors at work. To these I will turn in a few moments. In any case, one of the greatest needs of the psychology of the special senses at the moment is further controlled investigation of the exact mechanism, both physiological and psychological, of sensory adaptation. Till that work has been properly carried out, nobody has the right to be dogmatic, in any direction, concerning the results of continued periods of stimulation.

In the third place we are blamed for not

having produced sensational and remarkable effects. No reasonably controlled experiments on the effects of noise on work have so far done this. But take the small effects which we, and other investigators, have demonstrated, multiply them by the millions of times they occur in the case of the millions of people who are daily submitted to the type and level of noise that have been used in the experiments. Then, even if we agree that the ill-results are incident to change rather than to persistence of noise, the total effect at once becomes a matter of social significance. There is no need to overstate the case against noise.

3. An Industrial Experiment

We come now to what is practically the only really well-conducted direct investigation into the general effects of noise on work carried out upon industrial workers under industrial conditions. This also is due to the Medical Research Council of this country. There are, of course, numerous other reports about noise in industry, but all of them seem to assume that noise has detrimental effects either upon the worker, or upon his work, or on both; and they straightway devote themselves to a study of how noise can be reduced. H. C. Weston and S. Adams, however, investigated the direct effects of noise in a weaving shed (26). They surveyed the actual output of the operatives under ordinary conditions of work and with ear defenders. The level of noise reported under ordinary conditions was about 96 db., that is to say, we are here getting very near the range of intensity which may temporarily impair hearing. With the ear defenders the noise was reduced to some 85 db., so that it was still considerable, although the reduction was readily appreciable. Weston and Adams found that reduction of noise was accompanied by an hourly average increase of output for a group of ten skilled weavers, during an experimental period of twenty-six weeks, of between 1 and 2 per cent. Five of the weavers were definitely of the opinion that the new conditions were preferable, four thought they made no difference or were against their use, and one wished for further experiment. The increase of output noticed was greatest during the early stages of the day's work, and on the basis of this the investigators suggest that "even after years of work in a noisy environment, the weaver does not become completely adapted, or acclimatised to noise, but goes through the process of adaptation daily"*.

Once more we find the same general trend of results. Controlled investigation, whether in the laboratory or in the workshop, yields no evidence of any drastic impairment of function due directly to noise. Yet, with remarkable unanimity, all the researches show evidence of slight fall in the quality and quantity of work during the early stages of exposure to what may be called normal noise. There is general agreement, also, that this effect is not a function of the noise itself, but of the noise in relation to the type of work demanded. The more complex the task, and especially the more demands it makes upon the higher mental processes, the more marked are the effects of noise likely to be. Broadly

* Weston has continued his experiment in the weaving shed, and in twelve months of continuous controlled observation, he has confirmed the conclusions drawn from the first investigation.

speaking, we can say that, except where it is present in overwhelming amount, noise is harmful mainly as a distraction. It operates in the same manner as numerous other distractions, with the proviso that it certainly seems to be a stimulus to which the normal person can become rapidly and at least for the time being, completely adapted. As a distraction, however, while its effects do not justify the sensational statements that are often made, it is certainly harmful enough to provide a justification for all the efforts that can be made towards its reduction.

4. NOISE AND "NERVES"

Now what about the effects of noise on the person? At once our attitude must change from the more rigorous point of view of the experimentalist to that of the clinician. I will begin, in the accredited manner, by citing cases. The three instances I will take are, I need hardly say, in no way the products of my own imagination.

Here is a woman near, or just beyond, middle life. She is married to a man for whom she has a sincere regard, but who is physically rather weak, ineffective, unsatisfying. This worries her, and already she has become worked up into what she calls a "state of nerves". Now into the house next door come a newly married couple, young, magnificently vital, devoted. The house builder and the architect have, as usual, conspired to produce houses which transmit sounds with startling efficiency. She can hear everything that goes on next door, morning, noon and night. A few days, and her state has become noticeably worse. A few days more, and she has developed a fury of complaints against her neighbours. Complaints of what? Of the slamming of doors; nothing else.

Here is a man, young, normal, healthy. He sets himself to achieve high academic honours. He goes to a town in Germany, attaches himself to a University, works excessively, has late nights, sleeps poorly. His rooms are near a convent. Every morning, at an unhappily early hour, the booming convent bell wakes him up, and he cannot sleep again. After a while, he moves to more distant rooms. The bell no longer disturbs his sleep. But for months he can never hear its tolling, at any hour of the day, without a feeling of helpless fury.

Here is an old lady, desperately poor, unbendingly independent. She will fend for herself till she is dead. There are many things she would dearly like and cannot have. Her neighbours, not bad ones by any means, have many of these things. They instal a loud-speaker. The old lady hates and loathes the loud-speaker. She loses no chance of campaigning bitterly against "this disgraceful noise".

The first of these three instances is already over the border line of mental sanity. The second and the third are not. In all three, noise is made the burden of complaint that fundamentally has other and deeper causes as well. It is indeed astonishing how rarely we meet complaints against noise, how often against noises. Where objection is to noises we can suspect, with good reason, a psychological and special history of the trouble. This is the case again and again, how often only a statistical survey which has not been carried out could reveal. A man tired, run down, bored, maladjusted, uninterested, seizes upon anything outstanding from his environment to explain to himself and to others the unsuccess that life has brought him.

Noise, as we have seen, is just one of the things which, by its qualities, stands out prominently on almost any background. So it is not too much to say that whenever, in any community, a sweeping and passionate condemnation of noise is popular, there, within that community, are almost certainly a lot of people who are ill-adjusted, worried, attempting too much, or too little. The complaint against noise is a sign, sometimes, of a deeper social distress.

This, I am convinced, is the truth. But we must not let it carry us too far. It would be wrong to say that therefore noise in itself does not seriously matter. It may be admitted that, in a large number of instances, if noises were cut out or reduced, the people who object to them would soon transfer their complaints to some other outstanding feature of the environment. But of course nobody ever has argued, or ever would, that this would occur always, or be the only effect to follow. Every investigator has shown that noise has irritating and disturbing qualities. We know more about these than we did. We know how they are specifically connected with loudness, with pitch, with discontinuity, with lack of directional quality, with the meaning of the noise in relation to whatever task is demanded. No matter what reasons originally make noise, or some specific noise, a primary irritant, once it is on the way to becoming that, its presence undiminished will aggravate, deepen and perhaps establish the trouble. In another connection we saw how intense noise may help to develop and establish a deafness which is originally traceable to causes that had nothing to do with noise—to the after-effects of diseases like scarlet fever, or measles, or to the results of respiratory difficulties. So with many of the nervous effects of noise. The sounds complained of may or may not have originated the difficulties experienced, but in either case these are not likely to be alleviated permanently unless the noise can be reduced.

There is one particularly interesting reason which may, perhaps, tend to tie up noise with nervous complaints. Very intense sounds, sounds of high pitch and sounds whose source cannot readily be found or seen are all primary fear stimuli. In their presence most young children shrink and show the behaviour characteristic of fear or of timidity, and at some time or other, when sounds possessing these characters have occurred, most adults must have felt the unmistakable thrill of fright. Fear, of all human reactions, is probably the one most likely to get tied up with nervous disorders. So it is perhaps not too highly speculative to use this as a partial explanation of the frequency of those complaints against noise which have a nervous basis.

5. IRRITATION AND WORK

It has been very frequently observed that noise is extremely likely to set up irritation, especially in the early periods of exposure, or when the noise itself is intermittent. It is also true that, under experimental conditions, irritation, even to an exaggerated degree, is perfectly consistent with a high level of immediate performance. But once again we must not allow this observation, though it is absolutely sure so far as it goes, to carry us too far. Very possibly irritation may have more important effects upon remote than upon immediate performance. This is, at any rate, a common belief. A man who has been thrown into a mood of irritation may nevertheless do his immediate task very well. But when he turns away to something else in which he is perhaps a little less interested, or which, for any reason whatever, he treats as less obviously important, the mood may still persist, and may affect adversely whatever he attempts to do.

Here we approach one aspect of a problem which experimental psychologists have investigated very fully of late years. The general tendency of any attitude which has been set up by repeated or continued experience to hang on when its immediate conditions have been removed has been given the technical name of "perseveration". A man, for example, learns to drive a motor-car and becomes thoroughly familiar with the positions of all the controls. He now transfers to a different make of car in which the controls are not in the same relative positions. He may apparently adapt himself successfully; but, especially if he becomes a little tired, or out of sorts, or if his attention is disturbed, he may sometimes revert to the old movements, perhaps with unfortunate results. There seems no doubt that some persons are more liable to this tendency towards perseveration than others.

Now, for some reason difficult to appreciate, experimental psychologists have elected to investigate perseveration almost solely by a study of simple motor or simple intellectual processes. The general results have been confusing, and this was only to be expected. The course of life teaches very nearly everybody to adapt himself rapidly to a rapidly changing world, and all the senses which have to do with the interpretation of stimuli from the outside world get daily practice in keeping time with changes in those stimuli. But if we could study mental attitudes which depend largely upon stimuli coming from inside the organism, very likely we should find that these attitudes would persist, in despite of many changes of outward stimulation. The mood of the morning may run through the whole day, colouring everything that happens.

Experiments on moods are not easy to carry out, for moods are difficult to produce to order and to control, once they have been established. But the problems of technique are not insoluble. Miss C. Kendrew, for instance, has recently hit upon a method which seems likely to yield some highly interesting results. The investigation is in its early stages, and so far it has concerned young children only, but it may well turn out to have a significant bearing upon one of the pressing problems of noise.

Children are selected who, on a basis of general observation, seem likely to be relatively easily elated or depressed. They are given an interesting competitive task, and the conditions are so arranged that certain of them must fail and hence that they will not get a desired reward. Now these children are taken and are given an experimental task, which has nothing whatever to do with the original competition. The results so far are certainly striking. Some children fall upon their new task with unusual vigour, as if to compensate for their recent defeat and disappointment. But others consistently are slower, less accurate than usual. Irritation, disappointment, defeat, have spread over to something which had nothing to do with their initial establishment. Whether the one effect or the other is observed, seems to depend upon temperamental differences, the precise order of which has yet to be made out.

It is hazardous, of course, straightway to apply the results of these preliminary experiments upon young children to the case of the irritation produced in adults by noise. But in certain important respects the situations are parallel, and it may be that, while the irritating effects of noise produce little or no apparent deterioration of immediate work, they will be found to have an unhappy influence upon more remote performance and behaviour. If this can be shown to be the case, it is clear that the force of the objections against noise would at once be immensely strengthened.

I have stated the evidence, though briefly and with inadequate discussion. In special cases because of its physiological cost, in many more because of its effects on work—effects which are perhaps small individually but industrially and socially mount up into great significance—in general because it can have a profound psychological influence, sometimes initiating, more often aggravating, individual and social unhappiness, the case for serious and organised attempts to reduce noise is overwhelming.

Chapter III

1. WHAT CAN BE DONE?

We have now seen that, although the case against noise is often exaggerated, it rests upon a firm scientific foundation. There remains the question, no doubt to most people by far the most interesting and important, of what measures of defence should be adopted. It is hard to resist the conclusion that, while many people clamour for noise reduction, few indeed have ever thought very seriously about the difficulties that must be met. Some of these are highly technical ones, and spring from the fact that sound, like love, "laughs at locksmiths", and has a most disconcerting habit of finding its way through or across all manner of defences. Others are directly practical, and arise from the fact that many of the noises which produce the greatest crop of justifiable complaints are made by devices which, in other ways, are considered indispensable to advancing civilisation, or have their irritating qualities because they have to act as warnings in times of crisis. Yet much could be done, much ought to be done, and it seems desirable that I should attempt to indicate at least some of the general lines along which defences against noise could most reasonably be built.

When any nuisance threatens to become public, or already affects life in general, modern society always has two lines of possible defence. We may provide armour for the individual, or restrictions for the generality. The contemporary cricketer, for instance, going in to bat against a very fast bowler, may march to the wickets encased in vast pads up to his thighs, or on occasion may even don a kind of mail breastplate. But at the same time his friends—and particularly, perhaps, his older friends—may write to the newspapers in the hope that the ruling authorities may, in time, make certain methods of cricketing attack illegal. So with noise: we may encourage the individual to put on ear defenders, and we may urge governing powers to impose restrictive regulations. Let us first consider what may be called private defences.

2. EAR DEFENDERS

If the ordinary individual wants to shut out noise he probably stuffs wads of dry cotton-wool into his ears. This is not altogether unlike trying to dam up a tolerably vigorous stream of water with a stretched tennis-net. He would do better, however, if he saturated the wool with glycerine, carefully kneading the material to expel all possible air bubbles; or if he used vaseline, working it into the wool until a uniform mass was obtained. Both of the last two are, in fact, about as good ear defenders of a very simple type as can be made, particularly when protection is needed from intense explosive or detonating sounds. Wax cones, of the kind sometimes used by swimmers for other purposes, and employed in gunnery work in the Italian Navy, are also useful, and there is a type of ear defender, familiarly known during the Great War as "Tommy", which has come well out of severe tests. "Tommy" consists of a hollow soft rubber spherical bulb, with an opening on one side surrounded by a rubber flange. The external diameter of the bulb is about 9·5 mm., and of the flange 11·5 mm.

Another device which has fairly well satisfied the most stringent experimental tests is the one used by Weston and Adams in the investigation to which I referred in the second chapter. This is known as the Mallock-Armstrong ear defender. It consists of a small instrument, bulb-shaped at one end to fit into the ear. At the other end is a delicate membrane of gold-beaters' skin, not very tightly stretched, and enclosed between two discs of fine mesh wire gauze. The inner gauze disc is so placed that it prevents the membrane from vibrating with great amplitude and so diminishes the apparent loudness of the sounds that get through. As the membrane possesses a low natural frequency, components of noise of high pitch probably get transmitted less readily than lower tones.

Most of the serious experimental research which has been done on ear defenders concerns the protection of the ear from injury that may be caused by intense explosions at a short distance. This, however, presents a type of specialised problem of comparatively rare occurrence in daily life. Usually we desire to cut out, or to reduce, certain noises, but to leave other sounds to be transmitted as readily as ever. Within limits this can be done in a great many cases, but the apparatus required is neither cheap nor easily manageable, and for these and other reasons the ear defender must probably always remain the resource of the specialist, and can hardly become a part of the regular equipment of the ordinary person. Nobody could be expected to plug in an ear protector every time he goes for a ride in a tube train, or a noisy tram, or when he passes a riveting machine at work, or a pneumatic drill. This type of defence is for the workman in a heavy industrial plant; for the gunner; for the quarryman; for the signaller at work near heavy artillery. In all these cases, and others like them, where there is danger of physiological lesion due to noise stimulation, some proved ear defender ought to be used. Even so, it would be worth while carrying out much more research upon the production of a type of ear protector especially adapted for use in the continuous yet changing noise of heavy machine rooms in modern factories; for most of the existing devices fall far short of the ideal [22].

While in many ways the enormous increase of

easy and rapid transport facilities has added
new problems of noise to life, in other obvious
ways it has increased the chance of relief. So the
city-dweller, who is supposed to be jaded with
din, is often told to protect himself by taking car,
or bicycle, tram, bus, or train for spells of quiet
in the country. This is good advice, no doubt;
yet perhaps the country should not be too remote, or the spells of rest for the ear very long.
I have known plenty of ardent city lovers who
have soon become as heartily tired of silence as
ever they were of noise, and who have returned
to their familiar sounds with outward grumbling
and with inward content. And I have known
many more who seem to find it necessary to take
not a few of their noises to the country with
them. But here, maybe, we approach the problems of defence by public restriction.

3. DEFENCE AGAINST OUTSIDE AIR-BORNE SOUND

There is an increasing popular demand for laws
and by-laws against noise. Before anybody
lightly joins this, he should try to make a sort
of census of the kinds of noises against which

complaints are most frequent and most bitter. The majority of them are of the intermittent, or sporadic, type which in fact occupy only a relatively small part of daily life, except in special cases or areas. They are, for instance, noises like the clanging of bells; the sounds of heavy traffic during the night or early morning; motor-bicycles; motor-horns; various noises incidental to modern methods of building, such as those made by riveting machines, or cement mixers; the pneumatic drill. There are, of course, some apparent, or perhaps real, exceptions outside of the heavy workshop. More and more objections are raised every year against the continuous roar of traffic in busy towns and on crowded roads; but most people agree that the greatest difficulty comes with the occasional sounds that ring through the background, rather than with the background itself. Since the primary function of most of these occasional sounds is to compel attention, no matter how great may be the noise of the background, comparatively little can be done with them by way of general regulation. This applies, for example, to motor-horns, which must be penetrating and thrusting; and, since they are meant for use in very varying background conditions, it is necessary also to allow them a fairly large margin of safety in respect of loudness and pitch.

Popular and to some extent official opinion, however, has fastened upon the motor-horn nuisance as one that can be dealt with immediately and drastically by legislation. In many cities and in some countries there are already silence areas and silence periods in and during which a motor driver is prohibited from using his horn. The legislation is obviously experimental, but its immediate result has been of considerable interest from a psychological point of view. I think in practically every case, after such legislation has been put into force, it has been justified popularly, not so much on the ground of the reduction of noise effected, but because it has been followed by no increase, and sometimes by a decrease in the number of serious road accidents. It is always assumed that the absence of a warning signal is the cause of the decrease of accidents, because it makes a driver proceed more cautiously, or because it makes other people who may be on the road

67

more alert. This may be true, though it has not yet been proved. It is further assumed that the initial effect will be lasting. This is a great deal more doubtful. Any psychologist could off-hand quote many well-attested cases in which the introduction of novel conditions has produced marked effects for a brief period, but, as acclimatisation has set in, the initial effects have disappeared. Thus the driver who is kept from using his warning signal may proceed cautiously for a while, until he finds that he can go safely, and then, with no conscious purpose, may revert to an older and faster method. Only continued experiment can show whether this is likely to happen or not.

On the face of the matter, and always provided that the rules are so interpreted that the car driver is allowed some margin for the exercise of his own judgment, no particular difficulty seems likely to arise. But clearly, as in nearly every other case of noise, the problem is not one of noise alone. So far, except in very special cases, it is night driving only that is affected. In very brightly illuminated areas all may be well; and equally in areas where the only illumination is that from the head-lamps of cars on the road. The chief difficulties are likely to occur in areas in which the artificial illumination is indifferent, and here it seems safe enough to say that rules unintelligently administered may easily be worse than no general rules at all.

A matter that appears to receive very little consideration is the extraordinary variety of warning sounds that are used. Any manufacturer seems to be able to fit his car with any sort of horn he pleases, or any owner to alter his sound signals as he likes. But all the experiments show that occasional sounds are the more disturbing in proportion to their variety and their unexpected qualities. A greater uniformity, both as to loudness and as to pitch, in sounds used for warning signals in traffic is most highly desirable, and ought not to be particularly difficult to secure.

The motor-horn nuisance can no doubt be dealt with to some extent by general compulsory legislation. But, as in the case of most of the other noise nuisances of a public kind, its problems can ultimately be solved only by slow education and training in the exercise of thoughtfulness and good manners. A warning signal is an offensive weapon, not an indication of superior rights. It should be part of the routine of training to which all future drivers of cars will be submitted to see that this is fully appreciated.

As regards heavy traffic noises during the night or early morning, no doubt not a little could be achieved if both road and vehicle construction were considered more seriously than is usual from the point of view of reduction of sound. But the sensible line, so far as general regulation goes, would seem to be to take first the case of developing areas, and to see that town-planning and road construction are both attacked as parts of the same general problem. Heavy traffic routes should be planned so that as far as possible they are removed from areas of dwelling-house construction. Where important and thoroughly established markets are near fully populated districts which cannot be, or can only slowly be, depopulated or transformed into business areas, little beyond increased irritation can be produced by immediate restrictive legislation.

As everybody knows, the peacefulness of many small villages and country towns has been to some extent preserved or restored by the rapid development of the "by-pass" road movement. But much more might be done. A "by-pass" road should be that and nothing else. When, as frequently has been allowed to happen, houses spring up along the new roads, all the old problems are brought back again. Partly because of the unintelligent or inconsiderate use of building sites, and partly on account of the unavoidable prodigious increase of road traffic, "by-pass" roads tend to get so crowded that not a few motorists prefer the older ways. All this only shows again how every noise traffic problem is mixed up with others and cannot be sensibly treated in isolation.

Aeroplanes present certain noise problems which may be particularly difficult. The sounds emitted by aeroplane engines possess most of the qualities which make noises compelling and unpleasant. Consequently the assignment of sites for aeroplane factories and testing grounds and for aerodromes is a matter of great public interest, and should very definitely not be pro-

71

ceeded with in any haphazard manner. Foresight and planning are needed. That the noise of aeroplanes in flight may be extremely annoying and troublesome on occasion is perfectly true. We have seen already that sounds whose source is uncertain or unusual are always apt to attract attention, and most sounds from the air fall into this class. It is, however, very difficult to see what useful rules could be enforced other than those already enacted against low flying. All that is possible is to get whatever comfort one can from the fact that as air travel becomes more and more a part of the routine of daily life the noise of the aeroplane in flight will lose a considerable amount of its power to distract the attention and annoy the enforced listener.

No doubt the vigorous development of research into the most efficient type of engine-exhaust silencer will do much towards the reduction of unwanted sound wherever the internal combustion engine is used; although it is easy to envisage a state of affairs in which silently flying aeroplanes might bring with them particularly horrible consequences. The perforated baffle type of engine-exhaust-silencer, which is now, because it is simple and relatively cheap, much the most frequently used, may under certain circumstances be rather inefficient. As has been said, sound of high pitch is the most difficult to mask and also the most irritating, and the common type of silencer not infrequently lets high pitched components of a noise come through with ease. The best way to reduce some of the noise problems of the motor-bicycle, the motor-car, the aeroplane, the pneumatic drill, the riveting machine, the cement mixer would be to encourage even more active research into the production of really efficient silencing devices. Some of the modern methods of graphical sound analysis briefly described in the first chapter will prove of great value in aiding the production of the ideal silencer.

Most people will probably agree that of all the outside, air-borne, man-made sounds—excepting detonations as on the whole incidental to special circumstances—two of the most disquieting are the rattle of the pneumatic drill and the devastating chatter of the riveting machines used in the erection of the steel framework of the modern large building. In both cases silencing mechanisms could be vastly improved, but apart from this it is not very easy to see what could be done by general regulation. The questions involved are for the practical research engineer. Perhaps some day road materials may be found which not only meet all reasonable demands in the way of strength and durability, but which can be quietly broken up as often as town, borough and city councils desire. Perhaps also the use of bolts may, as now seems possible, altogether oust the riveting machine. If so, great gain will be won. In the meantime the steel framework building must be erected, and apparently roads must frequently be broken up; and nobody will deny that both must be done as speedily and as well as possible.

As for the many other miscellaneous sounds which arouse antagonism in special cases: bells, loud-speakers, whistling trains, factory hooters, street hawker's calls and farmyard cocks—some of them are welcomed by at least as many people as hate them; most of them are incidental; a few have a venerable history, and a very large number of them are the subject of individual idiosyncrasies which cannot reasonably be dealt with by general rule.

It is easy to claim public defence against public noises. But a candid examination will show that troublesome noises in the outside world, like a good many other nuisances, are exceptionally difficult to control by public rules, save in special cases and in a few special areas. The finally effective method here, as elsewhere, is by individual action; and this, perhaps, will best come as a slow growth of that kind of education which attempts to produce habits of consideration for the well-being of others.

It must in any case always be remembered that the experiments show conclusively that most normal people adapt themselves to a general continuous background of sound rapidly and efficiently. As I have repeatedly insisted, the harm comes mainly from sporadic noises thrust upon the general background. Now if the noise level of the general background, and with it that of the sporadic sound, is reduced in intensity, after all not very much may be gained. For the less loud occasional noise will now, on its new background, produce much the same result as the former louder sounds. The man who in the city needs a thunderstorm to waken him may, in the country, be roused by the patter of soft rain on his window-pane. Not too much must be expected from the reduction of the general background of noise in the modern civilised community.

4. THE TRANSMISSION AND ABSORPTION OF NOISE

The people who could take the most immediately practical steps towards noise reduction, if they would, are, not the politicians, but the builder, the builder's constructor and the architect. Beyond doubt it is the mental worker who suffers most from noise, provided for the moment we are not thinking of the sick and the overstrained. The type of mental work most affected, moreover, is generally done indoors. When people think of hygiene in reference to buildings, they usually mean physiological hygiene. So we must have houses, and especially offices and public institutions, which harbour the least possible amount of dirt. Far be it from me to speak disparagingly of all this. Yet the result does seem frequently to provide us structures with glass-hard walls, ceilings and floors; with steel window-frames and furniture; with rounded corners and rigid framework. From these we banish carpets and hangings as much as we are able, since these might harbour unswept dust. Or again, in a different type of building, and no doubt mainly for economic reasons, the thinnest kinds of partition are used without the slightest regard for their properties in the transmission of vibration and of noise. Thus we may have interiors that look bright and clean, but also that throw back all the sounds made inside them from every surface upon which they strike, and that transmit from all quarters the noises made outside. Increased general physiological well-being is always a gain, but if it is won at the cost of needless mental irritation it is obviously less of a gain than it might be.

The acoustic properties of a great many modern buildings are bad for most normal people; but what about the patients in public hospitals and wards? These are often especially unfit to withstand the assaults of noise, and, as everybody knows, many of the most sweeping complaints against excess of sound that are made are made in their name. Buildings for the sick are often startlingly noisy to a degree altogether unnecessary, were the principles of mental hygiene as fully known, or as seriously considered, as those of physiological health and cleanliness. There is no necessary conflict between these. Mental health and serenity and physical well-being go hand in hand. Yet, in many buildings for the sick, not even the most elementary precautions are taken to avoid reverberating sound, and the minds of patients are unnecessarily harassed in consequence.

Each year more and more facts become known with regard to sound transmission and absorption. It is already possible to predict, within limits, what amount of transmission will take place through light building materials, such as paper and board partitions. In the case of heavier panels the resonances of the material itself become very important, and much more investigation is required. But the methods by which such investigation can be carried out are being improved rapidly and only need more open and public appreciation and support to make yet more rapid strides forward. The relative advantages of single and double partitions in reducing air-borne sound are becoming more exactly known and specified. In the noisy office, ward, or room of the modern type, a considerable relief can be obtained merely by the introduction of soft floor coverings, and the absorbent effects of these can be still further increased by the use of an underfelt, more especially as regards sound of relatively low pitch. Wall or ceiling absorbents can be constructed of felt with a fabric covering which can easily be kept clean. The sound absorption is increased significantly if small holes are bored in the covering. Pads of slag wool will help to reduce acoustic reverberations, and produce very nearly as marked an effect even when they are covered with perforated metal sheets.

If only private individuals and public bodies could be induced to demand mental comfort with as much enthusiasm as they call for physiological cleanliness, it would be possible to go a long way to secure the first—in so far as it can be secured by good environmental conditions—

without in any way sacrificing the second. Architects, builders and builder's constructors would be forced to consider much more seriously than they do at present the known facts about acoustic transmission and reverberation, and scientists would be encouraged to push on with that research without which further progress is impossible [28].

5. SUMMARY

If, then, we survey the whole question of noise reduction and prevention from a practical point of view, it appears that at present, at least, special cases apart, far more is to be done by private education than by public restriction. The craftsman, at work in heavy noise, should be encouraged to use the most efficient ear defenders. Persons who have to sound warning noises should learn to have greater thought for others, and, whenever possible, should regulate their warnings more or less in accordance with variations in the general sound background. The builder's manufacturer, the builder and the architect should all be stimulated to keep in touch with contemporary research on sound transmission and reverberation. Everything possible should be done to build up a strong general opinion that good conditions for the mental life are as important as healthy physiological conditions. Possibly more stringent regulative action is already desirable with regard to the use of the most efficient exhaust silencers for different types of engine, though the practical aspect of this matter may be rapidly changed as research proceeds, and in any case exhaust silencers are a defence against only a part of the noise due to the working of engines. Town-planning should go hand in hand with road construction and the prescription of heavy traffic routes. All the time, and fundamental for any genuine advance, scientific research should be warmly encouraged. It is unfortunate that the most difficult noises to reduce, and especially to cut out, are usually those of high frequency, and that these are at the same time among the most irritating and disturbing. When everything is done that reasonably can be done, we shall still certainly have to reckon with plenty of individuals who will object strenuously to noises that remain. These individuals, however, for the most part, present psychological problems which have to be dealt with, so far as that is possible, by other measures altogether.

LIST OF REFERENCES

Note. The following list of references is merely meant to help any person who may be interested to find further information concerning some of the points dealt with in this book. It should be borne in mind that most of the matters discussed are the subject of active investigation, and that knowledge regarding them is all the time accumulating. The list is itself highly selected and could have been greatly extended without difficulty. Most of the articles and books referred to, however, themselves contain many other references which the reader who wishes to make a more complete survey can consult.

(1) MYERS, C. S., *Text-Book of Experimental Psychology*, 1, p. 25. Cambridge University Press, 1925.

(2) See e.g. DAVIS, H. and SAUL, L. J., 'The frequency of impulses in the auditory pathways' (*Amer. J. Physiol.* 1932, 101, pp. 28 ff.); HALLPIKE, C. S. and RAWDON SMITH, A. F., 'The Helmholtz resonance theory of hearing' (*Nature*, 1934, 133, p. 614).

(3) A general account of some methods used in the measurement of noise is given by KAYE, G. W. C., in 'Noise and its measurement' (*Nature*, 1931, **131**, pp. 253–64).

(4) See e.g. OBATA, J. and MORITA, S., 'On the accuracy of the aural method of measuring noise' (*J. Acoust. Soc. Amer.* 1932, **4**, pp. 129–37).

(5) GALT, R. H., *J. Acoust. Soc. Amer.* 1929, **1**, pp. 147 ff.

(6) DAVIS, A. H., 'The measurement of noise' (*The Physical Society, Report of a Discussion on Audition*, Cambridge University Press, 1931, pp. 82–91).

(7) For a very clear account see a 'Note on the decibel' in SHAXBY, J. H. and GAGE, F. H., *Med. Res. Council, Special Rep.* **166**, pp. 30–2.

(8) RAWDON SMITH, A. F., 'Auditory fatigue' (*Brit. J. Psychol.* General Section, 1934, **25**, p. 1).

(9) See e.g. TROINA, F., 'Sulle alterazioni dell' udito nel personale aeronavigante' (*Valsalva*, 1933, **9**, pp. 337–53).

(10) For the best recent account of high-frequency deafness see GUILD, S. R., 'Pathology of high-tone deafness' (*Bull. Johns Hopkins Hosp.* 1934, **54**, pp. 315–77).

(11) WITTMAACK, K, 'Über Schädigung des Gehörsorgans durch Schalleinwirkung' (*Z. f. Ohrenh.* 1907, pp. 37–80); YOSHII, U., 'Experi-

mentelle Untersuchungen über die Schädigung des Gehörorgans durch Schalleinwirkung' (*ibid.* 1909, pp. 201–57).

(12) LOSANOFF, N., 'On the professional traumatisation of the ears of industrial workers' (*Acta Otolar*, 1930, **14**, pp. 393–438).

(13) NYSSEN, R. and HELMSMOORTEL, J., Jr., 'L'influence des excitations auditives intenses sur la pression artérielle chez les normaux et chez les sourds labyrinthiques' (*J. neur. et de psychiat.* 1930, **30**, pp. 47–9).

(14) SMITH, E. L. and LAIRD, D. A., 'The loudness of auditory stimuli which affect stomach contractions in healthy human beings' (*J. Acoust. Soc. Amer.* 1930, **2**, pp. 94–8).

(15) HARMON, F. L., 'The effects of noise upon certain psychological and physiological processes' (*Arch. of Psychol.* 1933, **147**, pp. 81).

(16) FRANKEL, F. H., *Effects of Noise on the Hearing of Industrial Workers.* Albany, N.Y. Dept. Labor, 1930. Special Report **166**, pp. 42.

(17) See e.g. BANISTER, H., 'The basis of sound localisation' (*The Physical Society, Report of a Discussion on Audition*, Cambridge University Press, 1931, pp. 104–13). This paper contains a large number of references to the important literature. Also SHAXBY, J. H. and GAGE, F. H., *The Localisation of Sounds in the Median Plane*; and JAMES, H. E. O. and MASSEY, M. E., *Some*

85

Factors in Auditory Localisation, Medical Research Council, Special Report Series, No. **166**.

(18) TRENDELENBURG, F., 'Objective Measurement and Subjective Perception of Sound' (*The Physical Society, Report of a Discussion on Audition*, Cambridge University Press, 1931, pp. 92–100).

(19) A brief general account is given in BANISTER, H., 'Hearing, I', in *A Handbook of General Experimental Psychology*, Clark University Press, 1934, pp. 895–6. See also FLETCHER, HARVEY, 'Physical measurements of audition and their bearing on the theory of hearing' (*J. Franklin Inst.* 1923, **196**, pp. 289–326).

(20) LAIRD, D. A., 'Experiments on the Physiological cost of noise' (*J. Nat. Inst. Indust. Psychol.* 1928–9, **4**, pp. 251–8).

(21) VERNON, H. M. and WARNER, C. G., 'Objective and subjective tests for noise' (*The Personnel Journ.* 1932, **11**, pp. 141–9).

(22) FRYER, D. A., 'A genetic study of motivation under changing auditory situations' (*Brit. J. Psychol.* General Section, 1933–4, **24**, pp. 408–33 and 1934–5, **25**, pp. 140–69).

(23) HARBINSON, M. R. and BARTLETT, F. C., 'An investigation into the relation between discomfort and disability resulting from glaring light' (*Brit. J. Psychol.* General Section, 1933–4, **24**, pp. 313–19).

(24) LAIRD, D. A. and COYE, K., 'Psychological measurements of annoyance as related to pitch and intensity' (*J. Acoust. Soc. Amer.* 1929, 1, pp. 158–63).

(25) POLLOCK, K. G. and BARTLETT, F. C., 'Psychological experiments on the effects of noise' (*Indust. Health Res. Bd. Reports*, No. 65, pp. 1–37).

(26) WESTON, H. C. and ADAMS, S., 'The effects of noise on the performance of weavers' (*ibid.* pp. 38–62).

(27) See GUILD, S. R., 'War deafness and its protection,' and the further references given in this article (*J. Lab. and Clin. Med.* 1918–19, 4, pp. 153–80).

(28) The most important work of an experimental kind on various practical problems connected with noise reduction, so far as this country is concerned, is being carried out at the National Physical Laboratory. Summaries of the work done and detailed references appear in the Annual Reports published for the Department of Sci. and Indust. Research by H.M. Stationery Office. Also see WAGNER, K. W., 'Geräusch und Lärm' (*Sitzber. d. preuss. Akad. Wiss.* 1931, 9, pp. 154–65).

CAMBRIDGE: PRINTED BY
W. LEWIS, M.A.
AT THE UNIVERSITY PRESS

For EU product safety concerns, contact us at Calle de José Abascal, 56–1°, 28003 Madrid, Spain or eugpsr@cambridge.org.

www.ingramcontent.com/pod-product-compliance
Ingram Content Group UK Ltd.
Pitfield, Milton Keynes, MK11 3LW, UK
UKHW010851090126
466816UK00011B/158